SPOTLIGHT ON GLOBAL ISSUES

TECHNOLOGIES THAT HELP THE PLANET

Therese M. Shea

ROSEN PUBLISHING

NEW YORK

Published in 2022 by The Rosen Publishing Group, Inc.
29 East 21st Street, New York, NY 10010

First Edition

Editor: Theresa Emminizer
Book Design: Michael Flynn

Photo Credits: Cover Hien Phung Thu/Shutterstock.com; (series globe background) photastic/Shutterstock.com; p. 4 Maridav/Shutterstock.com; p. 5 Barcroft Media/Getty Images; p. 6 Nicole Patience/Shutterstock.com; p. 7 Bettmann/Getty Images; pp. 9, 13 (all) Bloomberg/Getty Images; pp. 10, 11, 15, 27 picture alliance/Getty Images; p. 12 oraziopuccio/Shutterstock.com; p. 16 anatoliy_gleb/Shutterstock.com; p. 17 Hannah Peters/Getty Images; p. 18 Alessandro Pierpaoli/Shutterstock.com; p. 19 monicaodo/Shutterstock.com; p. 20 Mohamed Abdulraheem/Shutterstock.com; p. 21 Ramin Talaie/Corbis Historical/Getty Images; p. 23 (all) Peter Parks/AFP/Getty Images; p. 24 Riccardo Mayer/Shutterstock.com; p. 25 John B. Carnett/Popular Science/Getty Images; p. 26 Lano Lan/Shutterstock.com; p. 29 SOPA Images/LightRocket/Getty Images.

Cataloging-in-Publication Data

Names: Shea, Therese M.
Title: Technologies that help the planet / Therese M Shea.
Description: New York : Rosen Publishing, 2022. | Series: Spotlight on global issues | Includes glossary and index.
Identifiers: ISBN 9781725323643 (pbk.) | ISBN 9781725323674 (library bound) | ISBN 9781725323650 (6 pack)
Subjects: LCSH: Technology--Juvenile literature.
Classification: LCC T48.S54 2022 | DDC 600--dc23

Manufactured in the United States of America

CPSIA Compliance Information: Batch #CSR22. For further information contact Rosen Publishing, New York, New York at 1-800-237-9932.

CONTENTS

CHAPTER ONE

EARTH IN TROUBLE

"Climate crisis!" "Environmental disaster!" If you've been paying attention to the news at all, you've heard these phrases. It's hard not to be worried about the future of Earth—and the future of all life on our planet. People have polluted the air, water, and surface of Earth with harmful substances. They've used vast amounts of nonrenewable resources, such as the fossil fuels coal, oil, and natural gas, at an astonishing pace. These fuels, when burned to provide heat and energy, contribute to greenhouse gases in Earth's atmosphere. These gases trap the sun's heat, leading to global warming, or an increase in Earth's surface temperature. Global warming, in turn, is changing climates around the world. Climate change means melting ice sheets, rising sea levels, disappearing habitats, the extinction of animals and plants, extreme weather events, and threats to all species on Earth.

Governments and people are demanding changes to improve the future of Earth in the face of climate change.

While a study of Earth reveals that heating and cooling trends are natural over the planet's history, most scientists agree that the current rate of change is 10 times faster than in a normal period. They also agree that human activity is the cause. The situation is quite serious, but people have been doing things to help the planet. Some scientists, engineers, and even kids have been finding ways to tackle many of the problems that face Earth today. There's hope for the future, thanks to human ingenuity! Read on to learn about the inspiring technologies that are aiding Earth.

When you think of technology, you probably think of digital devices or machines with many moving parts. However, technology is any way that people use scientific knowledge to change or control their environment. So, technology can be tools, machines, or even a process of doing something. Technology isn't just a modern concept either. When early people crafted axes to cut down trees, these were tools of technology. When people later learned to farm, this was a process of technology. The construction of houses also relies on technology.

The more modern concept of technology began during the Industrial Revolution. This started in the 1700s in Great Britain, when the increased demand for goods led to many developments in manufacturing. People invented engines to power machines. New factories, which could produce goods more quickly, opened. The Industrial Revolution spread to other countries, including the United States. Soon, there were developments in transportation. Engines that powered factory machinery, vehicles, and other inventions used fuels, especially coal and oil. At the time, people mistakenly thought these resources were unlimited. They also didn't realize the dangerous environmental impact of the pollution that fossil fuels produced.

James Watt's improved steam engine spurred the Industrial Revolution. Steam moved its parts, which in turn could move machine parts.

As time went on, people began to comprehend the negative impact of fossil fuel consumption. New technology made it possible to produce electricity with sunlight and wind power. Hybrid and electric vehicles cut down on greenhouse gas emissions. People also realized the importance of protecting our natural resources and reducing water, land, and air pollution. Recent technological breakthroughs are building on these positive movements.

CHAPTER TWO

CARBON CAPTURE

Collecting and cleaning up certain kinds of pollution seems easier than doing so with others. We can see trash in landfills, for example. But what about the gases that are accumulating in our atmosphere and contributing to global warming and climate change? One of these gases is carbon dioxide. Human activities such as the burning of fossil fuels and the clearing of forests have increased the carbon dioxide in our atmosphere.

One of the most promising technologies to help combat the carbon dioxide problem is called carbon capture. Carbon capture involves removing carbon dioxide gas from waste air coming out of factories, power plants, and other industrial facilities. The gas can be removed in several ways. It can be absorbed by a solvent. It can be joined with the chemical amine. It can also be cooled and compressed to separate it from steam. The carbon dioxide gas that's captured can then be stored underground or even turned into a kind of fuel.

Some facilities have been built solely for direct air capture, which means pulling the carbon dioxide out of the air around us. Again, this carbon dioxide could be stored so it can't escape into the atmosphere. It also could be turned into a low-carbon fuel. It's also been pumped into greenhouses (since plants use carbon dioxide for photosynthesis). Scientists don't think carbon capture technology can be used to completely solve the problem of global warming, but it can help as humans make other changes to reduce overall carbon dioxide emissions.

This direct air carbon capture facility in British Columbia, Canada, uses fans to draw in air. The carbon dioxide in that air binds to a special filter so it can be removed.

CHAPTER THREE

ALTERNATIVE AIRPLANES

To combat greenhouse gas emissions, some people focus on driving less. They may think about flying less too. A round-trip flight from New York to California produces 20 percent of the greenhouse gases that the typical car produces in a whole year. The number of airplanes in the sky is expected to double by 2040, and these planes are also expected to fly more. So, we can only expect more greenhouse gases, right? Wrong! Airlines have been combating airplane emissions in recent years. Some are using cleaner biofuels rather than air-polluting jet fuel. Some biofuels cut greenhouse gas output by more than 50 percent. People who fly may also buy a carbon offset when they travel. This means they contribute to an organization or activity that's doing something to take carbon dioxide out of the atmosphere, such as wind farms or tree-planting services.

HY4 COCKPIT

Currently, the HY4 seats four people, but larger no-emission planes are slated for the future.

A few engineers are planning to change airplanes themselves. A plane called the HY4 carries a motor powered by fuel cells, which generate electricity from a mix of hydrogen and oxygen. A battery helps with takeoff. This plane's only emission is water vapor. The HY4 was born out of a partnership between aircraft manufacturer Pipistrel, fuel cell specialists Hydrogenics, Germany's University of Ulm, and the German Aerospace Center's Institute of Engineering and Thermodynamics. The researchers plan to study how high this plane can go, how varying temperatures affect its performance, and if they can increase its current range of 1,200 miles (1,931.2 km).

CHAPTER FOUR

PLANT-BASED FOOD

When we think of greenhouse gases, we often think of fossil fuel–burning factories and vehicles. Most people don't think of farms, but they should—at least livestock farms. According to the Food and Agriculture Organization of the United Nations, raising livestock contributes about 14.5 percent of the greenhouse gases that human activities cause. Cattle, which are raised for milk, meat, and other purposes, contribute the most greenhouse gases of any livestock. Besides carbon dioxide, cattle also add the gases methane and nitrous oxide to the atmosphere. While these gases don't last as long as carbon dioxide does in the atmosphere, they absorb more of the sun's heat.

About one-fourth of all land on Earth is used for livestock. Much of it is land needed for grazing. Sometimes this means land is cleared of forests, which are needed to absorb carbon dioxide. One-third of all farmland is used to grow food for livestock too. Livestock manure also adds to water pollution.

If more people ate plants rather than meat, it would help in the battle against global warming. Plant-based foods are a step in that direction.

Food technology could help with these problems. In recent years, businesses have been perfecting plant-based products that look and even taste like meat. One kind of plant-based burger uses pea and rice proteins, beet juice, oils, and other ingredients. The company that created the burger says it produces 90 percent less greenhouse gas emissions than a hamburger does. It also claims that if all Americans traded beef burgers for plant-based burgers, it would be the equivalent of taking 12 million fossil fuel–burning cars off the road for a year.

CHAPTER FIVE

SOLAR GLASS

If you've ever been in a big city, you've probably looked up at a skyscraper and marveled at its size. Imagine if every window in that skyscraper could absorb energy from the sun and convert it into electricity we can use. That's the idea behind solar glass.

Solar energy is a renewable source of energy, unlike nonrenewable sources such as coal, oil, and natural gas. Nonrenewable sources take millions of years to form in the ground and could run out one day. However, the sun's rays will never run out—at least not for billions of years! Solar energy also doesn't release pollution into the air as does the burning of fossil fuels for energy production.

Solar energy usually is collected by photovoltaic cells found within solar panels. When light particles hit a solar cell, the particles, called photons, knock electrons off the atoms of a semiconductor. These electrons produce a flow of electricity that can travel through wires to power devices or be collected for future use.

In solar glass, tiny solar cells—smaller than a grain of rice—are fixed in or on glass but work much the same as those in solar panels. The solar cells called Sphelar solar cells are placed on glass. They're reported to absorb more light than flat cells because they're round. According to predictions by a group at Michigan State University, if people installed solar glass in the window space that exists today, about 40 percent of U.S. energy needs could be fulfilled by solar glass.

So far, solar glass hasn't been as efficient as larger solar panels. It also doesn't allow as much sunlight to shine through. However, scientists have made major improvements.

CHAPTER SIX

EFFECTIVE ENERGY STORAGE

You might wonder why we don't just shut down electrical plants powered by coal, oil, or natural gas, which contribute to greenhouse gas emissions in the atmosphere. Solar and wind technologies can generate the electricity we need without creating polluting gases and particles, right? Unfortunately, the answer is "sometimes." The sun doesn't always shine and the wind doesn't always blow, so these energy sources must be used in combination with power plant electricity in places with varying weather patterns. In order to cut ties with traditional power plants altogether, we need a way to store the energy produced by solar and wind farms for future use. Engineers who work in the area of energy storage technology are hard at work coming up with a solution.

The Tesla Powerwall contains a lithium-ion rechargeable battery. When combined with solar panels, it can power a home off the electrical grid. A larger system would be needed to provide widespread power to the public.

One idea is a kind of rechargeable battery that could store a large amount of energy for an extended period of time. Currently, lithium batteries are used for some energy storage. However, these are very expensive to produce and maintain, and they would be even more so if used to store the energy needed for everyone in the general public. In fact, storing 12 hours of energy this way for the entire United States would cost $2.5 trillion, according to a study in the journal *Energy and Environmental Science*. So far, it's been challenging to find a battery that can both do the job and be cost effective, but many companies are trying to do just that.

Most of us don't think about electricity until the power goes out. The traditional U.S. power grid is an amazing engineering feat connecting electricity from hundreds of power plants to over 325 million people. The grid spans over 300,000 miles (482,803.2 km) of wires and uses transformers to both raise voltages for transmission and to lower voltages for safe delivery into people's homes. However, this grid, designed in the 1890s, isn't equipped to handle all the energy needs of today. The energy flows in one direction, from the power plants to the consumer, and for the most part, energy isn't stored. The current grid incorporates some renewable energy sources, such as solar and wind farms. However, because these aren't always available—such as during a lack of sun for solar farms and a lack of wind for wind farms—they haven't been a major part of the power supply yet. If people are to cut down on the use of fossil fuels in the creation of electricity, increasing electricity created by renewable energy sources is essential.

In 2018, solar power provided just 1.6 percent of the electrical generation in the United States and all renewables just 17 percent. This percentage could jump much higher with a new kind of electrical grid using the latest technology.

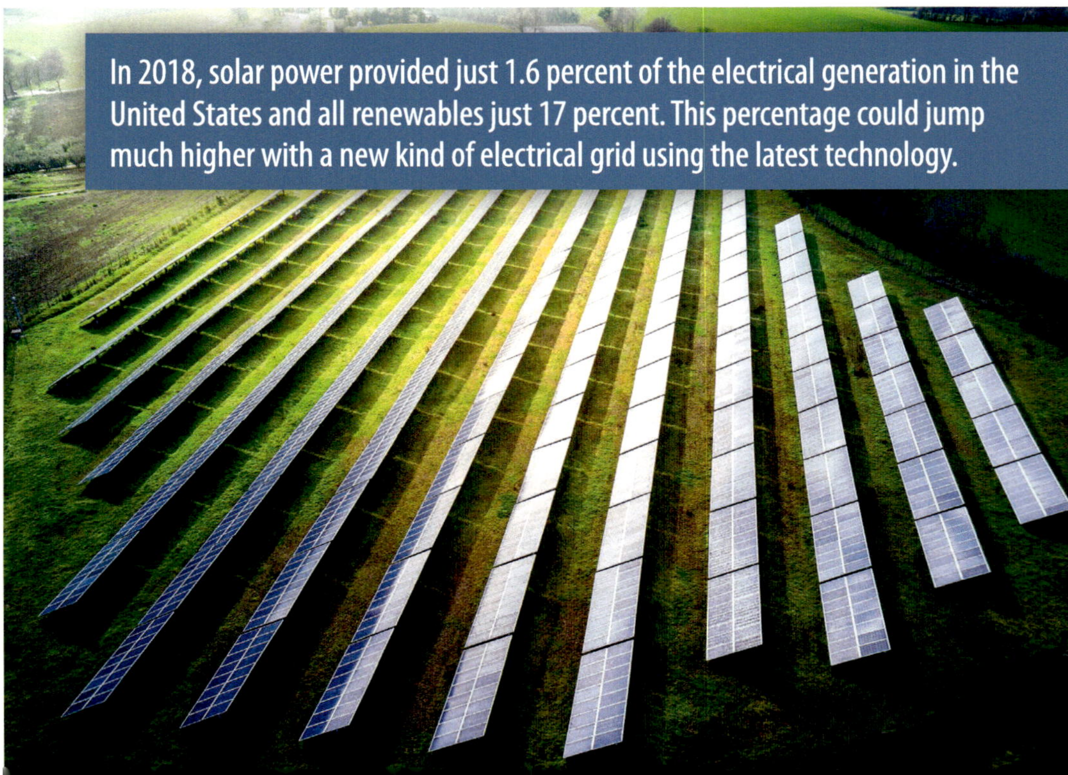

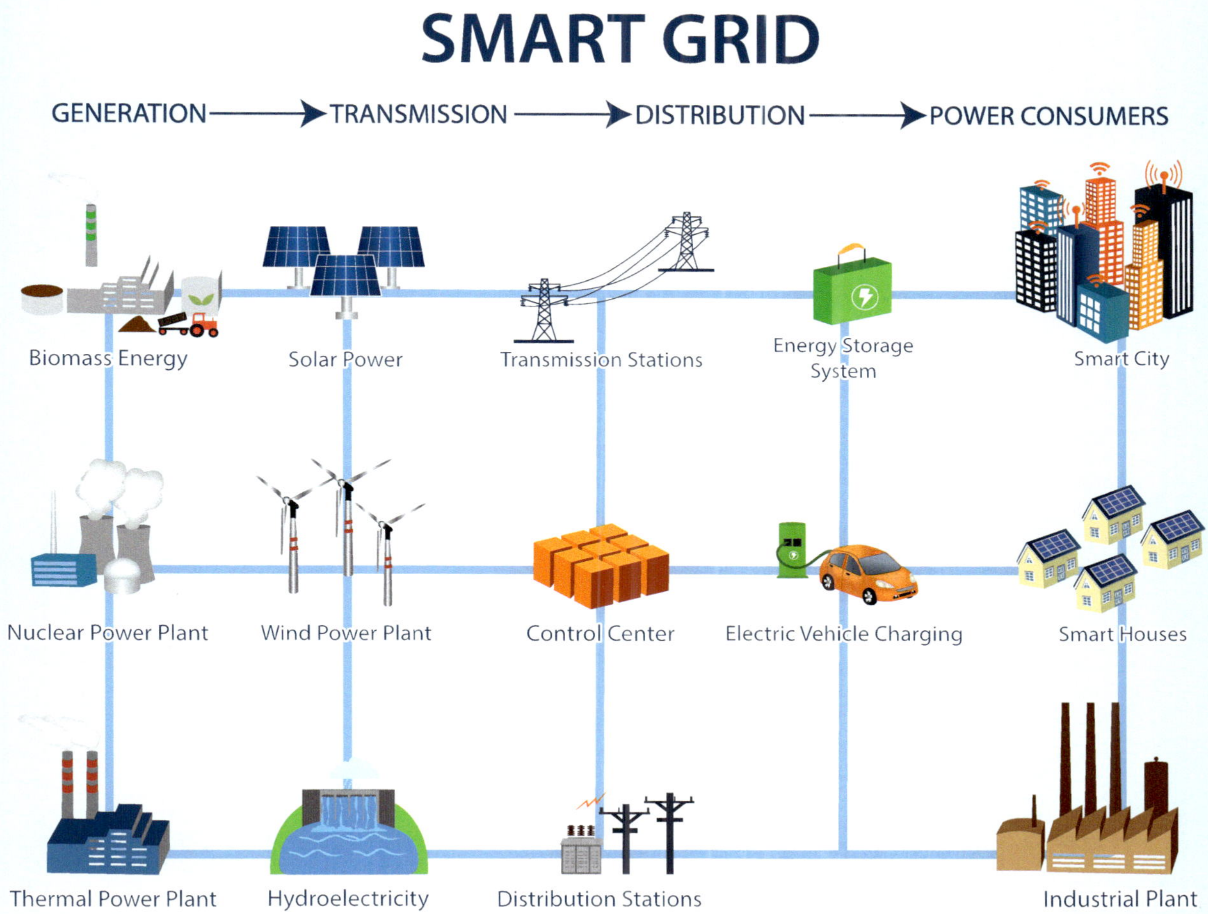

According to the Electric Power Research Institute, a power grid that uses renewable energy could reduce carbon dioxide emissions by more than 50 percent by 2030. The U.S. Department of Energy reports that a new kind of power grid, called a smart grid, will be built piece by piece over the next 10 years. It will consist of millions of parts, including computers, storage for renewable resources, and advanced technologies. It's predicted to transform society much the way the internet did.

CHAPTER SEVEN

PLASTIC FROM PLANTS

Plastic pollution is another problem facing Earth. Think of all the plastic items in your life: toys, TVs, containers, cups, bottles, bike and car parts, and more. Plastic is a strong, long-lasting, man-made material usually created from chemicals that can be shaped in many ways. Different kinds of plastics are made from different chemicals and used for different purposes. A big drawback is that most plastics aren't biodegradable, which means they don't break down easily. They may take hundreds of years to do so, in fact. This is a good thing until we want to throw the plastic object away. Plastic waste just builds up over time, becoming a major pollution problem that can harm wildlife. Plastics can also release chemicals that harm soil and water sources.

More than 90 percent of plastic waste isn't recycled. In recent years, more plant-based plastic products are coming into use.

People have long tried to make a kind of plastic that can biodegrade. Making the plastic strong enough to do its job is one concern. Making a plastic that doesn't break down too fast is another. However, inventors today are having breakthroughs in making plastics from plants or plant products such as corn and vegetable oils.

In 2014, a young man in Indonesia named Kevin Kumala got the idea to make plastic from an inexpensive vegetable called cassava. He launched a business producing plant-based plastic products. Not only do they break down after a few months, animals and people can safely ingest them! Plant-based plastics such as these may be the solution to our planet's plastic problem.

CHAPTER EIGHT

CLEAN WATER FOR ALL

Researchers estimate that millions of tons of trash end up in the oceans each year, and much of that is plastic. It's estimated that it takes about 450 years for a plastic bottle to break down in a landfill, but no one knows how long that takes in the ocean. Studies about plastic waste in the ocean have just begun. An organization called Ocean Conservancy calculates that over 8.8 million tons (8 million mt) of plastic ends up in ocean waters annually. Some is dumped on purpose and some ends up there by accident. Common plastic products such as straws and plastic bags can threaten aquatic animals' health and may even kill them. Ocean Conservancy reports that 100 percent of sea turtles—every sea turtle that has been studied—have plastic in their body.

While millions of volunteers help clean up our beaches, what can we do about the plastic in the ocean already? Many inventors are at work creating devices to capture plastics big and small. One of these inventions is called the Seabin, which looks much like a floating garbage can. It sucks water in through its top, including the trash floating on the surface. Plastic collects in its bag, while the cleaned water returns to the ocean. The Seabin can also collect oil floating on the ocean's surface. This oil would do great damage to the health of marine creatures and seabirds. As of 2020, the more than 850 Seabins at work have collected over 1 million pounds (453,600 kg) of garbage!

One of the inventors of the Seabin is Australian surfer and boat builder Pete Ceglinski. He's shown here with the device.

SEABIN PROJECT
FOR CLEANER OCEANS

According to the United Nations and the World Health Organization, there are more than 2 billion people who don't have access to clean drinking water. Unclean water can carry diseases such as cholera and typhoid. Hundreds of thousands of children die each year because of a lack of clean water. Good hygiene, including clean water with which to wash, is one of the simplest and best ways to prevent the spread of disease. However, getting clean water remains a challenge in some poor areas of the world because of poverty, war, or distance from water-cleaning facilities. In recent years, people have come up with new ways to clean water, including devices with filters and processes to separate pollutants from water.

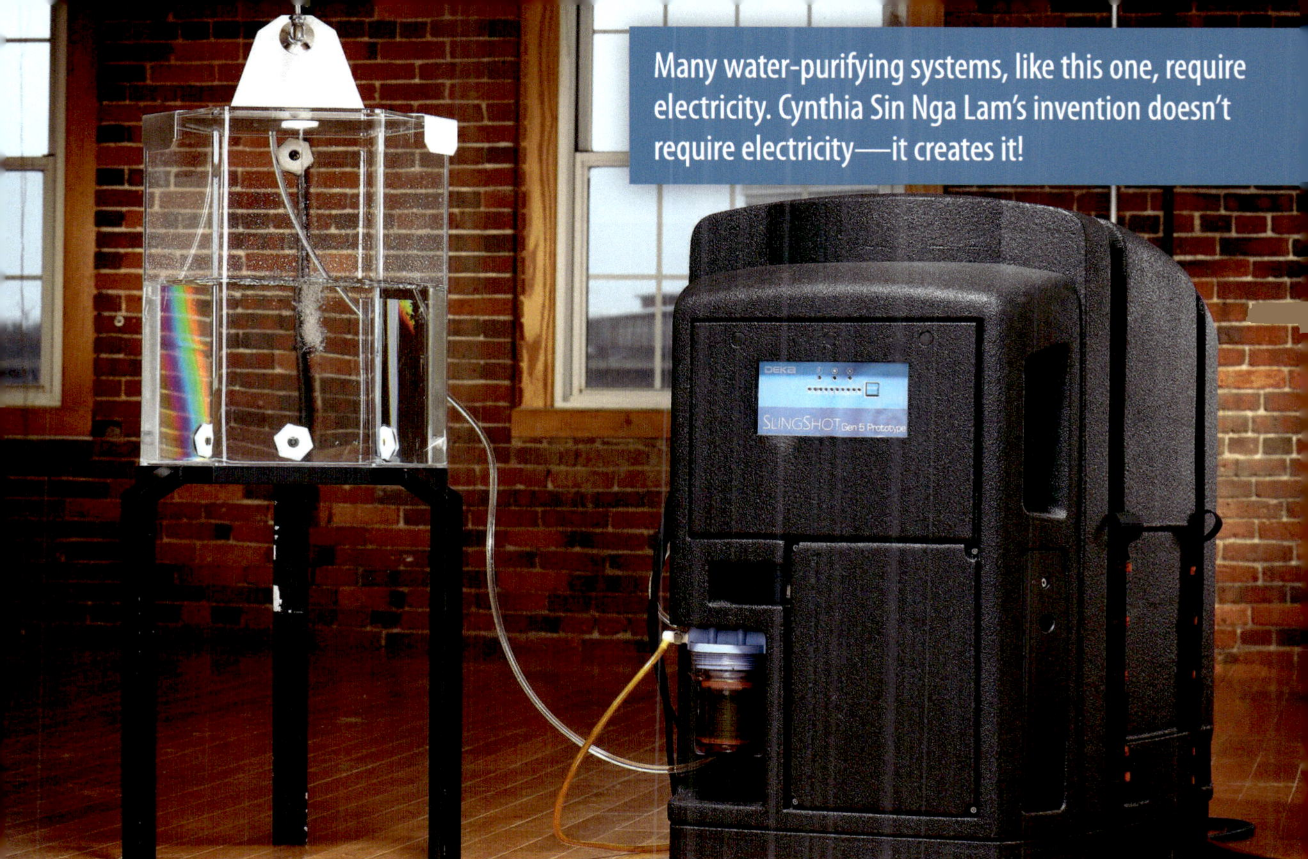

Many water-purifying systems, like this one, require electricity. Cynthia Sin Nga Lam's invention doesn't require electricity—it creates it!

An Australian teenager has invented a new way to clean water—as well as generate electricity. Cynthia Sin Nga Lam's device is called H2Pro. Dirty water goes in one end and a titanium mesh, powered by the sun, sterilizes the water before sending it through another filter. Sunlight speeds up a chemical process that produces hydrogen and oxygen, a reaction that could give power to a fuel cell, which can then provide electricity. Many regions that suffer from a lack of clean water also lack steady sources of electricity. This invention tackles two of the world's great problems. Another bonus is that the device would be affordable to many people. Since the introduction of this device, other scientists are working to perfect similar water-purifying and energy-generating devices.

CHAPTER NINE

VERTICAL FARMING

Technological advancements sometimes also focus on providing better medicine and making longer lives possible. However, that also means a growing population. It's thought that the world population may increase by 2 billion people by 2050. Yet, less and less arable land is available on which to grow food for people, and even land used for crops needs a rest sometimes to replenish the soil's nutrients. Technology has some answers—and one is called vertical farming.

Traditional farms have crops planted on one layer. Vertical farms are made up of many layers of planted crops that are stacked one on top of each other in a kind of a tower. With the help of natural and artificial lighting and temperature systems, vertical farms can be set up in vacant spaces such as old factories and warehouses. They can be established in cities, with the added advantage that crops don't have to be transported too far to get to an urban population.

A German company called Neofarms has designed a kind of vertical farm that people can use in their house. It's about the size of a refrigerator.

Vertical farming has several more benefits. One is that no pesticides are needed in such a controlled indoor environment. Every crop can be free of pesticides. Vertical farming also uses 70 to 95 percent less water than traditional farming. Additionally, crops can't be ruined by terrible storms, droughts, and other extreme weather events.

Scientists have discovered that a mix of red and blue LED light spurs photosynthesis in plants most efficiently. However, the great amount of energy needed to light large vertical farms is another challenge. People are studying these things to try to make these farms practical on a large scale.

CHAPTER TEN

DRONES CAN DO IT

People sometimes use drones, or unmanned flying vehicles, for military and for recreational reasons, but they're also increasingly used to help the planet. They can be fitted with cameras and sensors and flown to areas that people can't get to easily, such as polar regions or deserts. People can even drive drones into the ocean to monitor the health of water. Drones can fly over solar farms and wind farms to find damaged equipment. They can be used to observe animal populations in remote areas as well. While many drones run on batteries, solar-powered drones also exist.

In the future, we may need drones for another reason: pollination of flowers and plants. Many plants require pollen, which contains the plant's male reproductive cells, to be carried to the female part of another flower. It's believed that pollinators are required for about three-quarters of all plants grown for food, including fruits, nuts, and vegetables. Pollinators include bees, beetles, butterflies, and ants—and bees are among the most important of these. Unfortunately, the population of honeybees has been in decline, with beekeepers reporting huge losses each year. Pesticides, habitat loss, and parasites are the likely causes. So, how will crops and flowers be pollinated if we lose more pollinators? While it's possible to pollinate some crops by hand, that's a very difficult and time-consuming task on a large farm. Scientists in Japan have been developing small drones to carry pollen from one plant to the next. A test flight has successfully pollinated a flower.

A drone inspects lettuce at a farm. Drone use on farms helps cut down on overuse of fertilizers, water, and other resources.

CHAPTER ELEVEN

DO YOUR PART

In the past, people used technology to advance their businesses, increase their comfort, and further their own interests in other ways. Some of this technology had unintended harmful effects on our planet. However, many people are now using technology to help our planet. Still, we can't expect it to replace nature. We also can't rely on a technological breakthrough to rewind all the damage that's been done.

We all must take steps in our lives to help our planet. We can create achievable goals such as reducing our use of plastics, decreasing our use of nonrenewable resources, and supporting sustainability efforts. We can also educate ourselves about what we consume and about lifestyle changes to help the environment. If everyone did these things, we could make a real impact on the health of the planet. Technology can help—but we need to help too.

Young people are at the forefront of the modern environmental movement. Teens like Cynthia Sin Nga Lam are leading the way to a brighter, more sustainable future for the planet and its people. You don't have to invent new technology to help.

Activists are people who act strongly in support of a cause. Greta Thunberg is a Swedish teen activist who has gained international recognition for her work calling on world leaders to address climate change.

Whether you create a new solution to an environmental problem, stand up for the planet by using your voice, or simply make changes in your own life, your actions have the power to create a better future for our world.

GLOSSARY

arable (AA-ruh-buhl) Able to be farmed.

biofuel (BY-oh-fyool) A renewable fuel such as wood, composed of or produced from living matter.

cholera (KAH-luh-ruh) A deadly disease often caused by a bacterium and marked by stomach and intestinal problems.

fossil fuel (FAH-suhl FYOOL) A fuel—such as coal, oil, or natural gas—that is formed in the earth from dead plants or animals.

hybrid (HY-bruhd) Describing something that has two kinds of components that perform the same function.

hygiene (HY-jeen) Ways to keep yourself and your surroundings clean in order to maintain good health.

ingenuity (ihn-juh-NOO-uh-tee) Skill or cleverness that allows people to solve problems and create inventions.

LED (EHL-EE-DEE) Stands for "light-emitting diode," a semiconductor that gives off light when a current runs through it.

photosynthesis (foh-toh-SIHN-thuh-suhs) The process by which a plant turns water and carbon dioxide into food when the plant is exposed to sunlight.

photovoltaic (foh-toh-vahl-TAY-ihk) Having to do with the production of an electric current when two substances are exposed to light.

semiconductor (seh-mee-kuhn-DUHK-tuhr) Matter that allows some electricity or heat to move through it.

solvent (SAHL-vuhnt) A liquid used to dissolve another substance.

sustainability (suh-stay-nuh-BIH-luh-tee) Using methods that do not completely use up or destroy natural resources.

titanium (ty-TAY-nee-uhm) A strong and light silvery metal.

typhoid (TY-foyd) A deadly disease caused by a bacterium and marked by fever, headache, and intestinal problems.

voltage (VOLH-tihj) The force that would drive an electric current, measured in volts.

INDEX

PRIMARY SOURCE LIST

Page 5
Extinction rebellion protest outside the Bank of England in London. Photograph. Barcroft Media. April 25, 2019. Barcroft Media via Getty Images.

Page 7
James Watt 1788 Double-Action Sun-and-Planet Steam Engine. Photograph. Science & Society Picture Library / Contributor. SSPL via Getty Images.

Page 23
Pete Ceglinski, CEO and co-founder of Seabin Project. Photograph. PETER PARKS/AFP via Getty Images. June 08, 2018. AFP via Getty Images.

WEBSITES

Due to the changing nature of Internet links, Rosen Publishing has developed an online list of websites related to the subject of this book. This site is updated regularly. Please use this link to access the list: www.powerkidslinks.com/SOGI/technologies